REAL SCIENCE-4-KIDS



KOGS-4-KIDS

# PHYSICS CONNECTS TO LANGUAGE

Rebecca W. Keller, Ph.D.

Cover design: Rebecca W. Keller, Ph.D.
Opening page: Rebecca W. Keller, Ph.D.
Illustrations: Rebecca W. Keller, Ph.D.



Physics Connects To Language
ISBN # 978-1-936114-99-3

Published by Gravitas Publications, Inc.
www.gravitaspublications.com

Printed in United States

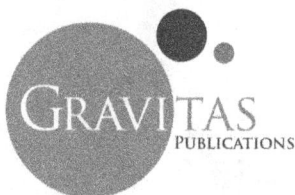

# Contents

# Introduction

## I.1 The language of science

Have you ever noticed that scientists use all kinds of fancy words like nucleosynthesis (nü-klē-ō-sin′-thə-səs) or photoelectric photometry (fō-tō-i-lek′-trik fō-tä′-mə-trē)? Many of the words that scientists use are long and difficult to pronounce. However, these words have been carefully selected by scientists as they put the field of science into verbal language. Each scientific word means a particular thing. There is a *language* to science.

## I.2 Latin and Greek roots

If someone had to memorize all of the words that scientists use, it would be a difficult task. Most of the words that are used come to us from two languages: Latin and Greek. Many of the words you will encounter in science will have Latin or Greek *word roots.* A word root is that part of a word that is derived from another word. For example, the word "biology" comes from the Greek word *bios*, which means "life," and *logy* which means "study of," so biology means "the study of life." The word tree illustration

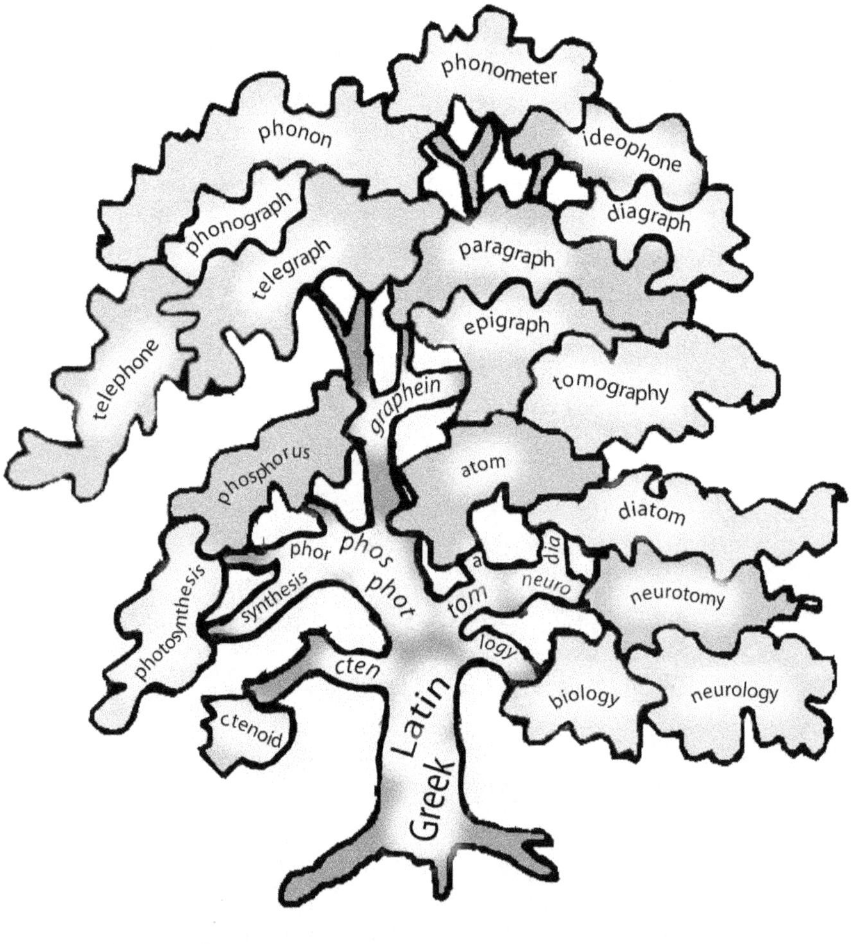

shows several different words and their Latin or Greek word roots. You can see that the words on the branches are the word roots, and those on the leaves come from these roots. In fact, many of the languages that people speak have Latin or Greek word roots. English, Spanish, Portuguese, German, and even Romanian all have some words that are similar because some words in each of these languages come from Latin or Greek. For example, the English, Spanish, Portuguese, French, and Italian words for school all come from the Latin word *schola.*

| "school" | Language |
|---|---|
| **schola** | **Latin** |
| escuela | Spanish |
| escola | Portuguese |
| scuola | Italian |
| school | English |
| ecole | French |

We can learn a lot about languages by learning the Latin and Greek word roots.

## I.3 How this book works

This workbook is called a "connection" because it *connects* the discipline of science to the language of science. Using this workbook will help you to more easily understand the different science subjects because you will know more about the language of science.

In *Find the Root,* the first section of each chapter, you will find the word root in a set of six English words. The pronunciation of each word is shown, and there is a pronunciation key at the end of the book.

For example, your word list may look something like this:

**deflate** (di-flāt′)

**inflate** (in-flāt′)

**flabellum** (flə-bel′-əm)

**flavor** (flā′-vər)

**conflate** (kən-flāt′)

**afflate** (ə-flāt′)

The word root can be three letters long, four letters long, or even five letters long. You are asked to look for the three to five letters (the cluster) that are common in each word. This is the **word root**. For example, this list of words has the following three letters in common:

**deflate**

**inflate**

**flabellum**

**flavor**

**conflate**

**afflate**

We see that all of the words have the common cluster of letters "f," "l," and "a" which make up the word root "**fla**."

The exercises in the first section of each chapter are designed to get you thinking about the words in the list and their common word root. You are asked to *guess* the meaning of the word root and any of the words in the list.

In *Learn the Root,* the second section of each chapter, the meaning of the word root is defined. For example, we found that the word root for this list is the three letter word root **fla**.

| | |
|---|---|
| de**fla**te | **fla**vor |
| in**fla**te | con**fla**te |
| **fla**bellum | af**fla**te |

In this example we discover that the meaning of the word root **fla** comes from the Latin word *flare* which means "wind" or "to blow." All of the words in the list have something to do with "wind" or "blowing." In this section you will be encouraged to try to define the words in the list *before* you look at the definitions. Guessing is good! It gets you thinking, even if your guesses are wrong.

The third section, *Definitions,* defines each of the words in the list. These definitions are taken from a number of different dictionaries including *Webster's Unabridged New Twentieth Century Dictionary 1972,* the *American Dictionary of the English Language 1828,* and a *Thesaurus of Word Roots of the English Language.* Additional Latin or Greek word roots for prefixes or suffixes are also given.

The meanings of the words in our example are:

**deflate**
1. to collapse by letting out air or gas
2. to lessen the importance of, as with money

(*de*—opposite)

**inflate**
1. to blow full of air, to expand
2. to raise the spirits of
3. to increase, raise beyond normal

(*in*—in)

**flabellum**
1. a large fan, usually carried by the Pope
2. in zoology or botany, a fan-shaped part or structure

**flavor**
an odor, smell, or aroma *carried by the wind*

**conflate**
1. blow together, bring together, collect
2. to combine, melt, fuse, or join

(*con*—together)

**afflate**
to blow or breathe upon
(*af*—to, toward)

In the next section, *Mix and Match*, you are given the opportunity to match the definitions of the words to the word list. For example:

| | |
|---|---|
| **deflate** | to blow or breathe upon |
| **inflate** | a large fan or fan-shaped structure |
| **flabellum** | an odor or aroma |
| **flavor** | to let out gas by collapsing |
| **conflate** | to expand |
| **afflate** | to blow together |

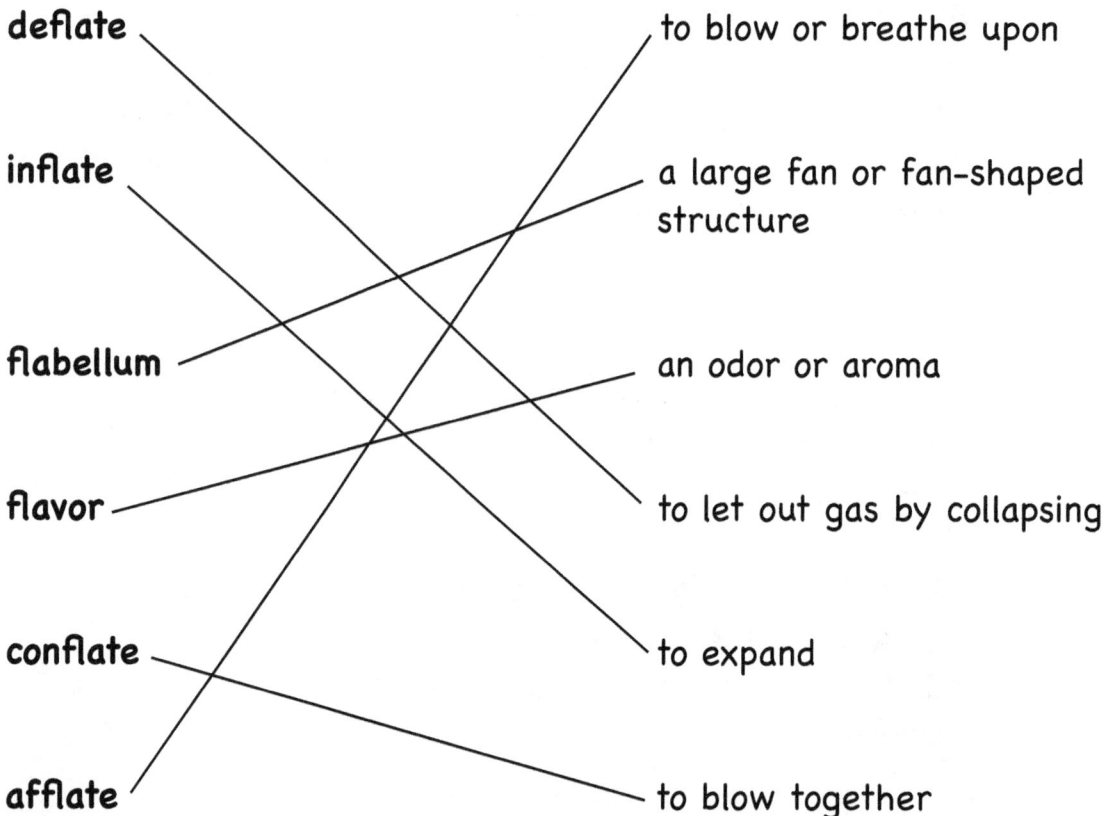

Now that you have learned the word root and the definitions of the different words in the list, you can use the *Test Yourself* section to see how well you remember them.

Finally, in the last section of each chapter, *Using New Words,* you will have a chance to make up a story or several sentences using the words you have learned.

For example:

*While carrying a rather **deflated** balloon down the sidewalk, and his hand **conflated** with his mother's, little Johnny was losing his interest with his mother's conversation with Mr. Longs, who awkwardly carrying his **flabellum**, was nevertheless chatting whole-heartedly. The **flavor** of his chocolate sundae was wearing off and his patience was at an end. **Afflating** hard upon his mother, which was his favorite way of gaining her attention, his mother made her farewells and Mr. Longs, still awkwardly carrying his **flabellum**, moved on. Johnny was saved from the boring conversation at last.*

written by Christopher Keller, age 9

Now that you have learned about how language connects to science, use this workbook as you learn Physics Level I, and most importantly...

Have fun!

# Chapter 1: Physics

## 1.1 Find the root

Look at the following words:

**physics** (fi′-ziks)

**apophysis** (ə-pä′-fə-səs)

**metaphysics** (me′-tə-fi-ziks)

**physiology** (fi-zē-ä′-lə-jē)

**diaphysis** (dī-a′-fə-səs)

**physique** (fə-zēk′)

There is a cluster, or group, of letters that is exactly the same in all six words. Can you find the cluster?

Circle the cluster that is the same in each word. Write the four letters that make up the cluster. ________ ________ ________ ________

Now look at the words carefully. Because they all have a common cluster, they all have meanings with some similarities.

Can you guess the meaning of the cluster?

______________________________________________

______________________________________________

Can you write a definition for any of the words in the list?

______________________________________________

______________________________________________

______________________________________________

## 1.2 Learn the root

**Word Cluster**

| | |
|---|---|
| **phys**ics | apo**phys**is |
| meta**phys**ics | **phys**iology |
| dia**phys**is | **phys**ique |

All of the words above have a common word root—**phys**. The word root **phys** comes from the Greek word *physis*, which means "to exist," "to be," or "to grow." All of the words have something to do with "existing," "being," or "growing."

Knowing that the cluster **phys** comes from the Greek word *physis*, try to guess the meanings of the words before looking at the definitions in the next section.

**physics** ____________________________________________

**apophysis** ____________________________________________

**metaphysics** ____________________________________________

**physiology** ____________________________________________

**diaphysis** ____________________________________________

**physique** ____________________________________________

## 1.3 Definitions

**physics** — the field of science concerned with properties, changes, and interactions of matter

**apophysis** — a natural process or outgrowth; in geology, an outgrowth of igneous rock
(*apo*—away from)

**metaphysics** — a branch of philosophy that deals with the nature of being or reality (ontology), the structure and origin of the world (cosmology), and the theory of knowledge (epistomology)
(*meta*—later, behind)

**physiology** — the branch of biology that deals with how living organisms and their organs and parts function
(*logos*—study of)

**diaphysis** — in anatomy, the central part of the bone between the growing ends
(*dia*—through)

**physique** — the structure, form, and appearance of the body

## 1.4 Mix and match

Draw lines to connect the words with their meanings.

| | |
|---|---|
| **physics** | the structure and form of the body |
| **apophysis** | a branch of biology dealing with the organs and parts of living organisms |
| **metaphysics** | the field of science dealing with the properties and interactions of matter |
| **physiology** | an outgrowth |
| **diaphysis** | the central part of a bone between the growing ends |
| **physique** | a branch of philosophy dealing with the nature of reality, the origin of the world, and the theory of knowledge |

## 1.5 Test yourself

Write the meaning next to each word below.

**physics** ______________________________

**apophysis** ______________________________

**metaphysics** ______________________________

**physiology** ______________________________

**diaphysis** ______________________________

**physique** ______________________________

## Extra:

Can you guess the meanings of the following words?

**physiocrat** (*crat*—ruled by) (fi′-zē-ə-krat)

______________________________

**epiphysis** (*epi*—upon) (i-pi′-fe-səs)

______________________________

## 1.6 Using new words

Write several sentences using each new word you have learned.

# Chapter 2: Gravity

## 2.1 Find the root

Look at the following words:

**gravity** (gra′-və-tē)

**gravitate** (gra′-və-tāt)

**aggravate** (a′-grə-vāt)

**grave** (grāv)

**gravid** (gra′-vəd)

**gravamen** (gra-vā′-mən)

There is a cluster of letters that is exactly the same in all six words. Can you find the cluster?

Circle the cluster that is the same in each word. Write the four letters that make up the cluster. ________ ________ ________ ________

Now look at the words carefully. Because they all have a common cluster, they all have meanings with some similarities.

Can you guess the meaning of the cluster?

________________________________________

________________________________________

Can you write a definition for any of the words in the list?

________________________________________

________________________________________

________________________________________

## 2.2 Learn the root

**Word Cluster**

| | |
|---|---|
| **grav**ity | **grav**itate |
| ag**grav**ate | **grav**e |
| **grav**id | **grav**amen |

All of the words above have a common word root—**grav**. The word root **grav** comes from the Latin word *gravis,* which means "heavy." All of the words have something to do with "heavy."

Knowing that the cluster **grav** comes from the Latin word *gravis,* try to guess the meanings of the words before looking at the definitions in the next section.

**gravity** ______________________________

**gravitate** ______________________________

**aggravate** ______________________________

**grave** ______________________________

**gravid** ______________________________

**gravamen** ______________________________

## 2.3 Definitions

**gravity** in physics, the force that draws two bodies towards each other; weight, heaviness

**gravitate** to move according to the force of gravity; to be attracted to or tend to move toward something

**aggravate** to make heavy, worse, or severe; to exaggerate (*ad*—to)

**grave** *(adj.)* requiring serious thought, important, weighty; solemn, sedate
*(noun)* a hole in the ground used to bury a body

**gravid** pregnant, with child

**gravamen** a complaint or grievance

## 2.4 Mix and match

Draw lines to connect the words with their meanings.

| | |
|---|---|
| **gravity** | to be attracted to or move toward something |
| **gravitate** | pregnant |
| **aggravate** | in physics, the force that brings two bodies together |
| **grave** | a complaint or grievance |
| **gravid** | to make worse |
| **gravamen** | requiring serious thought |

## 2.5 Test yourself

Write the meaning next to each word below.

**gravity** ____________________

**gravitate** ____________________

**aggravate** ____________________

**grave** ____________________

**gravid** ____________________

**gravamen** ____________________

### Extra:

Can you guess the meanings of the following words?

**gravimeter** (*meter*—measure) (gra-vi′-mə-tər)

____________________

**gravigrade** (*gradus*—to walk) (gra′-vi-grād)

____________________

## 2.6 Using new words

Write several sentences using each new word you have learned.

# Chapter 3: Potential

## 3.1 Find the root

Look at the following words:

**potential** (pə-ten′-shəl)

**impotent** (im′-pə-tənt)

**potentate** (pō′-tən-tāt)

**despot** (des′-pət)

**omnipotent** (äm-ni′-pə-tənt)

**potency** (pō′-tən-sē)

There is a cluster of letters that is exactly the same in all six words. Can you find the cluster?

Circle the cluster that is the same in each word. Write the three letters that make up the cluster. ________ ________ ________

Now look at the words carefully. Because they all have a common cluster, they all have meanings with some similarities.

Can you guess the meaning of the cluster?

______________________________________________

______________________________________________

Can you write a definition for any of the words in the list?

______________________________________________

______________________________________________

______________________________________________

## 3.2 Learn the root

**Word Cluster**

**pot**ential　　im**pot**ent
**pot**entate　　des**pot**
omni**pot**ent　　**pot**ency

All of the words above have a common word root—**pot**. The word root **pot** comes from the Latin word *potis,* which means "able" or "power." All of the words have something to do with being "able" or "powerful."

Knowing that the cluster **pot** comes from the Latin word *potis,* try to guess the meanings of the words before looking at the definitions in the next section.

**potential** ____________________________________________

**impotent** ____________________________________________

**potentate** ____________________________________________

**despot** ____________________________________________

**omnipotent** ____________________________________________

**potency** ____________________________________________

## 3.3 Definitions

**potential** something that can, but has not yet, come into being; unrealized, undeveloped

**impotent** without power; lacking physical strength; ineffective, helpless
(*in[im]*)—without)

**potentate** a person who possesses great power; a ruler or monarch

**despot** master, lord; an absolute ruler; a king of unlimited power; tyrant

**omnipotent** all powerful; possessing unlimited power
(*omni*—all)

**potency** the state or quality of being potent or powerful

## 3.4 Mix and match

Draw lines to connect the words with their meanings.

| | |
|---|---|
| **potential** | a tyrant |
| **impotent** | something that can, but has not yet, come into being |
| **potentate** | the state of being powerful |
| **despot** | a monarch |
| **omnipotent** | without power |
| **potency** | all powerful |

## 3.5 Test yourself

Write the meaning next to each word below.

**potential** ______________________________

**impotent** ______________________________

**potentate** ______________________________

**despot** ______________________________

**omnipotent** ______________________________

**potency** ______________________________

### Extra:

Can you guess the meanings of the following words?

**plenipotentiary** (*pleni*—fully) (ple-nə-pə-ten′-shə-rē)

______________________________

**potentiometer** (*meter*—to measure) (pə-ten-shē-ä′-mə-tər)

______________________________

## 3.6 Using new words

Write several sentences using each new word you have learned.

# Chapter 4: Heliocentric

## 4.1 Find the root

Look at the following words:

**heliocentric** (hē-lē-ō-sen′-trik)

**parhelion** (pär-hēl′-yən)

**helium** (hē′-lē-əm)

**isohel** (ī′-sə-hel)

**Helios** (hē′-lē-əs)

**heliotropism** (hē-lē-ä′-trə-pi-zəm)

There is a cluster of letters that is exactly the same in all six words. Can you find the cluster?

Circle the cluster that is the same in each word. Write the three letters that make up the cluster. ________ ________ ________

Now look at the words carefully. Because they all have a common cluster, they all have meanings with some similarities.

Can you guess the meaning of the cluster?

______________________________________________

______________________________________________

Can you write a definition for any of the words in the list?

______________________________________________

______________________________________________

______________________________________________

## 4.2 Learn the root

### Word Cluster

**hel**iocentric

**hel**ium

**Hel**ios

par**hel**ion

iso**hel**

**hel**iotropism

All of the words above have a common word root—**hel**. The word root **hel** comes from the Greek word *helios,* which means "sun." All of the words have something to do with the "sun."

Knowing that the cluster **hel** comes from the Greek word *helios,* try to guess the meanings of the words before looking at the definitions in the next section.

**heliocentric** ______________________________

**parhelion** ______________________________

**helium** ______________________________

**isohel** ______________________________

**Helios** ______________________________

**heliotropism** ______________________________

## 4.3 Definitions

**heliocentric** literally, "sun-centered"; having or relating to the Sun as the center; calculated or viewed from the center of the Sun (*kentron*—center)

**parhelion** a fake sun; bright lights near the Sun that look like the Sun but are not—sometimes the lights are connected with one another by a white arc or halo

**helium** a gas that is the second element in the periodic table, has an atomic weight of 4.003, and is found in the Sun

**isohel** a line drawn on a map that connects two points that receive equal amounts of sunlight

**Helios** the Greek sun god

**heliotropism** the tendency of certain plants or other organisms to turn or bend in response to the direction of light from the Sun

## 4.4 Mix and match

Draw lines to connect the words with their meanings.

| | |
|---|---|
| **heliocentric** | the Greek sun god |
| **parhelion** | the tendency of a plant or other organism to be influenced by the Sun's path |
| **helium** | sun-centered |
| **isohel** | a gaseous element found in the Sun |
| **Helios** | a fake sun |
| **heliotropism** | a line on a map connecting two points that receive equal sunlight |

## 4.5 Test yourself

Write the meaning next to each word below.

**heliocentric** ______________________________

**parhelion** ______________________________

**helium** ______________________________

**isohel** ______________________________

**Helios** ______________________________

**heliotropism** ______________________________

### Extra:

Can you guess the meanings of the following words?

**helioscope** (*skopein*—view) (hē′-lē-ə-skōp)

______________________________

**heliochrome** (*chroma*—color) (hē′-lē-ə-krōm)

______________________________

## 4.6 Using new words

Write several sentences using each new word you have learned.

# Chapter 5: Nucleus

## 5.1 Find the root

Look at the following words:

**nucleus** (nü′-klē-əs)

**nucleoid** (nü′-klē-oid)

**mononucleosis** (mä-nō-nü-klē-ō′-səs)

**pronucleus** (prō-nü′-klē-əs)

**nucleoplasm** (nü′-klē-ə-pla-zəm)

**nucleoprotein** (nü-klē-ō-prō′-tēn)

There is a cluster of letters that is exactly the same in all six words. Can you find the cluster?

Circle the cluster that is the same in each word. Write the four letters that make up the cluster. ________ ________ ________ .________

Now look at the words carefully. Because they all have a common cluster, they all have meanings with some similarities.

Can you guess the meaning of the cluster?

______________________________________________

______________________________________________

Can you write a definition for any of the words in the list?

______________________________________________

______________________________________________

______________________________________________

## 5.2 Learn the root

**Word Cluster**

| | |
|---|---|
| **nucl**eus | **nucl**eoid |
| mono**nucl**eosis | pro**nucl**eus |
| **nucl**eoplasm | **nucl**eoprotein |

All of the words above have a common word root—**nucl**. The word root **nucl** comes from the Latin word *nux*, which means "nut" or "kernel." All of the words have something to do with "nut" or "kernel."

Knowing that the cluster **nucl** comes from the Latin word *nux*, try to guess the meanings of the words before looking at the definitions in the next section.

**nucleus** ________________________________________

**nucleoid** ________________________________________

**mononucleosis** ________________________________________

**pronucleus** ________________________________________

**nucleoplasm** ________________________________________

**nucleoprotein** ________________________________________

## 5.3 Definitions

**nucleus** a thing or part forming in the center from which something grows; in botany, a kernel or nut or seed; in chemistry and physics, the central part of an atom; in biology, an organelle in a eukaryotic cell

**nucleoid** in biology, a region in prokaryotic cells that is similar to the nucleus
(*oid*—similar to)

**mononucleosis** an infectious disease that creates an excessive number of blood cells called leukocytes that contain a single nucleus
(*mono*—one)

**pronucleus** in zoology, the nucleus of either the male or female reproductive cell (gamete) before it combines with another gamete to create a fertilized egg

**nucleoplasm** a plasm (fluid material) inside the nucleus of a cell

**nucleoprotein** a protein found inside the nucleus of a cell

## 5.4 Mix and match

Draw lines to connect the words with their meanings.

| | |
|---|---|
| **nucleus** | an infectious disease that causes an excessive number of leukocytes to be produced |
| **nucleoid** | a protein found in the nucleus of a cell |
| **mononucleosis** | a kernel or seed |
| **pronucleus** | the fluid material inside the nucleus of a cell |
| **nucleoplasm** | the nucleus of either a male or female gamete |
| **nucleoprotein** | a region similar to a nucleus that is found in prokaryotic cells |

## 5.5 Test yourself

Write the meaning next to each word below.

**nucleus** ______________________________

**nucleoid** ______________________________

**mononucleosis** ______________________________

**pronucleus** ______________________________

**nucleoplasm** ______________________________

**nucleoprotein** ______________________________

### Extra:

Can you guess the meanings of the following words?

**nucleonics** (*-nics*—having to do with electro*nics*) (nü-klē-ä'-niks)

______________________________

**nucleoform** (*forma*—form) (nü'-klē-ō-fôrm)

______________________________

## 5.6 Using new words

Write several sentences using each new word you have learned.

# Chapter 6: Electric

## 6.1 Find the root

Look at the following words:

**electric** (i-lek′-trik)

**electrode** (i-lek′-trōd)

**dielectric** (dī-ə-lek′-trik)

**electrolysis** (i-lek-trä′-lə-səs)

**pyroelectric** (pī-rō-i-lek′-trik)

**electrophilic** (i-lek-trə-fi′-lik)

There is a cluster of letters that is exactly the same in all six words. Can you find the cluster?

Circle the cluster that is the same in each word. Write the six letters that make up the cluster. _____ _____ _____ _____ _____ _____

Now look at the words carefully. Because they all have a common cluster, they all have meanings with some similarities.

Can you guess the meaning of the cluster?

______________________________________________________________

______________________________________________________________

Can you write a definition for any of the words in the list?

______________________________________________________________

______________________________________________________________

______________________________________________________________

## 6.2 Learn the root

**Word Cluster**

| | |
|---|---|
| **electr**ic | **electr**ode |
| di**electr**ic | **electr**olysis |
| pyro**electr**ic | **electr**ophilic |

All of the words above have a common word root—**electr**. The word root **electr** comes from the Greek word *electrum,* which means "shining" or "amber." The original meaning is uncertain but is related to electrons.

Knowing that the cluster **electr** comes from the Greek word *electrum,* try to guess the meanings of the words before looking at the definitions in the next section.

**electric** ______________________________

**electrode** ______________________________

**dielectric** ______________________________

**electrolysis** ______________________________

**pyroelectric** ______________________________

**electrophilic** ______________________________

## 6.3 Definitions

**electric** charged with or conveying electricity; producer of or produced by electricity; operated by electricity; (electricity is a form of energy in which electrons are moved from one place to another as in a wire)

**electrode** either of the two terminals, either positive or negative, in a battery
(*ode*—way, path)

**dielectric** an insulator such as glass or rubber that does not permit the flow of electrons
(*dia*—through, across)

**electrolysis** 1. in chemistry, the breakdown of a substance into ions by the action of an electric current passing through the substance

2. the use of an electric needle to remove unwanted hair by destroying the roots

(*lyein*—remove, loosen)

**pyroelectric** the development of electric polarity by a change in temperature
(*pyro*—fire)

**electrophilic** a chemical or compound that accepts electrons
(*phila*—love)

## 6.4 Mix and match

Draw lines to connect the words with their meanings.

| | |
|---|---|
| **electric** | a crystal that will develop electric polarity by means of heat |
| **electrode** | an insulator |
| **dielectric** | electron-loving |
| **electrolysis** | conveying electricity |
| **pyroelectric** | the positive or negative terminal of a battery |
| **electrophilic** | the breakdown of a compound into ions by an electric current |

## 6.5 Test yourself

Write the meaning next to each word below.

**electric** ______________________________

**electrode** ______________________________

**dielectric** ______________________________

**electrolysis** ______________________________

**pyroelectric** ______________________________

**electrophilic** ______________________________

### Extra:

Can you guess the meanings of the following words?

**photoelectric** (*photo*—light) (fō-tō-i-lek′-trik)

______________________________

**hydroelectric** (*hydro*—water) (hī-drō-i-lek′-trik)

______________________________

## 6.6 Using new words

Write several sentences using each new word you have learned.

# Chapter 7: Thermal

## 7.1 Find the root

Look at the following words:

**thermal** (thər′-məl)

**diathermy** (dī′-ə-thər-mē)

**exothermic** (ek-sō-thər′-mik)

**hypothermia** (hī-pə-thər′-mī-ə)

**thermometer** (thər-mä′-mə-tər)

**endothermic** (en-də-thər′-mik)

There is a cluster of letters that is exactly the same in all six words. Can you find the cluster?

Circle the cluster that is the same in each word. Write the five letters that make up the cluster. ______ ______ ______ ______ ______

Now look at the words carefully. Because they all have a common cluster, they all have meanings with some similarities.

Can you guess the meaning of the cluster?

____________________________________________

____________________________________________

Can you write a definition for any of the words in the list?

____________________________________________

____________________________________________

____________________________________________

## 7.2 Learn the root

**Word Cluster**

| | |
|---|---|
| **therm**al | dia**therm**y |
| exo**therm**ic | hypo**therm**ia |
| **therm**ometer | endo**therm**ic |

All of the words above have a common word root—**therm**. The word root **therm** comes from the Greek word *therme,* which means "heat." All of the words have something to do with "heat."

Knowing that the cluster **therm** comes from the Greek word *therme,* try to guess the meanings of the words before looking at the definitions in the next section.

**thermal** ______________________________

**diathermy** ______________________________

**exothermic** ______________________________

**hypothermia** ______________________________

**thermometer** ______________________________

**endothermic** ______________________________

## 7.3 Definitions

**thermal** having to do with heat; warm or hot; such as *thermal underwear, thermal capacity, thermal unit,* etc.

**diathermy** in medicine, a procedure where heat is produced in the tissues of the skin by an electric current (*dia*—through, across)

**exothermic** in chemistry, a chemical reaction that gives off heat (*exo*—out, outward)

**hypothermia** in medicine, a body temperature that is below normal (*hypo*—under)

**thermometer** an instrument that measures temperature (*meter*—to measure)

**endothermic** in chemistry, a reaction that consumes or takes in heat (*endo*—within, inner)

## 7.4 Mix and match

Draw lines to connect the words with their meanings.

| | |
|---|---|
| **thermal** | a body temperature below normal |
| **diathermy** | having to do with heat |
| **exothermic** | a chemical reaction that consumes heat |
| **hypothermia** | a chemical reaction that gives off heat |
| **thermometer** | in medicine, a procedure that produces heat in skin tissue |
| **endothermic** | an instrument that measures heat |

## 7.5 Test yourself

Write the meaning next to each word below.

**thermal** ______________________________

**diathermy** ______________________________

**exothermic** ______________________________

**hypothermia** ______________________________

**thermometer** ______________________________

**endothermic** ______________________________

### Extra:

Can you guess the meanings of the following words?

**homoiothermous** (*homoio*—same, similar) (hō-moi-ō-thər′-məs)

______________________________

**poikilothermous** (*poikilos*—various) (poi-kē-lə-thər′-məs)

______________________________

## 7.6 Using new words

Write several sentences using each new word you have learned.

# Chapter 8: Induction

## 8.1 Find the root

Look at the following words:

**induction** (in-dək′-shən)

**deduce** (di-düs′)

**reduce** (ri-düs′)

**product** (prä′-dəkt)

**aqueduct** (a′-kwə-dəkt)

**transducer** (tranz-dü′-sər)

There is a cluster of letters that is exactly the same in all six words. Can you find the cluster?

Circle the cluster that is the same in each word. Write the three letters that make up the cluster. ________ ________ ________

Now look at the words carefully. Because they all have a common cluster, they all have meanings with some similarities.

Can you guess the meaning of the cluster?

_______________________________________________

_______________________________________________

Can you write a definition for any of the words in the list?

_______________________________________________

_______________________________________________

_______________________________________________

## 8.2 Learn the root

**Word Cluster**

| | |
|---|---|
| in**duc**tion | de**duc**e |
| re**duc**e | pro**duc**t |
| aque**duc**t | trans**duc**er |

All of the words above have a common word root—**duc**. The word root **duc** comes from the Latin word *ducere,* which means "to lead" or "to pull." All of the words have something to do with "leading" or "pulling."

Knowing that the cluster **duc** comes from the Latin word *ducere,* try to guess the meanings of the words before looking at the definitions in the next section.

**induction** ______________________________

**deduce** ______________________________

**reduce** ______________________________

**product** ______________________________

**aqueduct** ______________________________

**transducer** ______________________________

## 8.3 Definitions

**induction** leading or bringing into; in physics, the act of some field or force bringing an electric or magnetic effect to something
(*in*—towards, within)

**deduce** to draw conclusions from premises; infer; gather
(*de*—from)

**reduce** to diminish, decrease, shorten; pull back
(*re*—back again)

**product** something that has been "led forward" by being produced or marketed; something that has been led to as a result of a set of conditions
(*pro*—forward, before)

**aqueduct** a channel leading water to a new place
(*aqua*—water)

**transducer** in physics, a device that leads power from one system through something to another system
(*trans*—through)

## 8.4 Mix and match

Draw lines to connect the words with their meanings.

| | |
|---|---|
| **induction** | a device that leads power from one place to another |
| **deduce** | a channel that leads water from one place to another |
| **reduce** | leading or bringing into |
| **product** | to draw conclusions from a set of premises |
| **aqueduct** | something that has been led forward by being produced or marketed |
| **transducer** | diminish, decrease |

## 8.5 Test yourself

Write the meaning next to each word below.

**induction** ______________________________

**deduce** ______________________________

**reduce** ______________________________

**product** ______________________________

**aqueduct** ______________________________

**transducer** ______________________________

### Extra:

Can you guess the meanings of the following words?

**educate** (*ex[d]*—out) (e′-jə-kāt)

______________________________

**viaduct** (*via*—road) (vī′-ə-dəkt)

______________________________

## 8.6 Using new words

Write several sentences using each new word you have learned.

# Chapter 9: Decibel

## 9.1 Find the root

Look at the following words:

**decibel** (de′-sə-bel)

**decanal** (də-kā′-nəl)

**duodecimal** (dü-ə-de′-sə-məl)

**hexadecane** (hek-sə-de′-kān)

**decimeter** (de′-sə-mē-tər)

**December** (di-sem′-bər)

There is a cluster of letters that is exactly the same in all six words. Can you find the cluster?

Circle the cluster that is the same in each word. Write the three letters that make up the cluster. ________ ________ ________

Now look at the words carefully. Because they all have a common cluster, they all have meanings with some similarities.

Can you guess the meaning of the cluster?

______________________________________________

______________________________________________

Can you write a definition for any of the words in the list?

______________________________________________

______________________________________________

______________________________________________

## 9.2 Learn the root

**Word Cluster**

**dec**ibel
**dec**anal
duo**dec**imal
hexa**dec**ane
**dec**imeter
**Dec**ember

All of the words above have a common word root—**dec**. The word root **dec** comes from the Latin word *decem,* which means "tenth" or "ten." All of the words have something to do with "ten."

Knowing that the cluster **dec** comes from the Latin word *decem,* try to guess the meanings of the words before looking at the definitions in the next section.

**decibel** ____________________

**decanal** ____________________

**duodecimal** ____________________

**hexadecane** ____________________

**decimeter** ____________________

**December** ____________________

## 9.3 Definitions

**decibel** a unit for measuring the volume of sound; ten decibels is equal to one bel

**decanal** pertaining to a dean; [*dean* was originally the term for a person in charge of ten monks, later soldiers]

**duodecimal** a system of numbering with twelve as its base [*duo*—two; *duo + dec*, two plus ten equals 12]

**hexadecane** a hydrocarbon with sixteen atoms of carbon (*hexa*—six; *hexa + dec*, six plus ten equals sixteen)

**decimeter** a unit of measure equal to one tenth of a meter (*meter*—to measure)

**December** the tenth month of the early Roman calendar; today the 12th month in the year

## 9.4 Mix and match

Draw lines to connect the words with their meanings.

| | |
|---|---|
| **decibel** | a dean |
| **decanal** | the tenth month of the Roman calendar |
| **duodecimal** | a hydrocarbon with sixteen carbon atoms |
| **hexadecane** | a numbering system with base twelve |
| **decimeter** | ten bels |
| **December** | one tenth of a meter |

## 9.5 Test yourself

Write the meaning next to each word below.

**decibel** ____________________

**decanal** ____________________

**duodecimal** ____________________

**hexadecane** ____________________

**decimeter** ____________________

**December** ____________________

## Extra:

Can you guess the meanings of the following words?

**dime** *(hidden root)*

____________________

**dozen** *(hidden root)*

____________________

## 9.6 Using new words

Write several sentences using each new word you have learned.

# Chapter 10: Petroleum

## 10.1 Find the root

Look at the following words:

**petroleum** (pə-trō′-lē-əm)

**petrify** (pe′-trə-fī)

**petrology** (pə-trä′-lə-jē)

**petroglyph** (pe′-trə-glif)

**petrous** (pe′-trəs)

**petrogenic** (pe-trə-je′-nik)

There is a cluster of letters that is exactly the same in all six words. Can you find the cluster?

Circle the cluster that is the same in each word. Write the four letters that make up the cluster. ________ ________ ________ ________

Now look at the words carefully. Because they all have a common cluster, they all have meanings with some similarities.

Can you guess the meaning of the cluster?

____________________________________________

____________________________________________

Can you write a definition for any of the words in the list?

____________________________________________

____________________________________________

____________________________________________

## 10.2 Learn the root

**Word Cluster**

| | |
|---|---|
| **petr**oleum | **petr**ify |
| **petr**ology | **petr**oglyph |
| **petr**ous | **petr**ogenic |

All of the words above have a common word root—**petr**. The word root **petr** comes from the Greek word *petra*, which means "rock." All of the words have something to do with "rock."

Knowing that the cluster **petr** comes from the Greek word *petra*, try to guess the meanings of the words before looking at the definitions in the next section.

**petroleum** ____________________

**petrify** ____________________

**petrology** ____________________

**petroglyph** ____________________

**petrous** ____________________

**petrogenic** ____________________

## 10.3 Definitions

**petroleum** a flammable, oily liquid that is found in layers of rock in the ground, is composed largely of hydrocarbons, and is refined for use as fuel
(*oleum*—oil)

**petrify** to convert organic matter, such as bone tissue or wood, into stone

**petrology** the study of the composition, structure, and origin of rocks
(*logy*—to study)

**petroglyph** any marking, inscription, or drawing etched into the face of a rock or cliff
(*glyph*—carving)

**petrous** of or like a rock; hard, stony

**petrogenic** something that produces rocks
(*gen*—producing)

## 10.4 Mix and match

Draw lines to connect the words with their meanings.

| | |
|---|---|
| **petroleum** | like a rock |
| **petrify** | the study of rocks |
| **petrology** | an oily liquid found in rocks that is used for fuel |
| **petroglyph** | to make into stone |
| **petrous** | rock producing |
| **petrogenic** | an etching or carved drawing found on a rock |

## 10.5 Test yourself

Write the meaning next to each word below.

**petroleum** ______________________________

**petrify** ______________________________

**petrology** ______________________________

**petroglyph** ______________________________

**petrous** ______________________________

**petrogenic** ______________________________

## Extra:

Can you guess the meanings of the following words?

**Peter** *(hidden root)*

______________________________

**lamprey** *(hidden root)* (lam′-prē)

______________________________

## 10.6 Using new words

Write several sentences using each new word you have learned.

# Pronunciation Key

| | |
|---|---|
| a | add |
| ā | race |
| ä | palm |
| â(r) | air |
| b | bat |
| ch | check |
| d | dog |
| e | end |
| ē | tree |
| f | fit |
| g | go |
| h | hope |
| i | it |
| ī | ice |
| j | joy |
| k | cool |
| l | love |
| m | move |
| n | nice |
| ng | sing |
| o | odd |
| ō | open |
| ô | jaw |
| oi | oil |
| oo | pool |
| p | pit |
| r | run |
| s | sea |
| sh | sure |
| t | take |
| u | up |
| ü | sue |
| v | vase |
| w | way |
| y | yarn |
| z | zebra |
| ə | a in above |
| | e in sicken |
| | i in possible |
| | o in melon |
| | u in circus |

# More REAL SCIENCE-4-KIDS Books
## by Rebecca W. Keller, PhD

**Building Blocks Series** yearlong study program — each Student Textbook has accompanying Laboratory Notebook, Teacher's Manual, Lesson Plan, Study Notebook, Quizzes, and Graphics Package

Exploring Science Book K (Activity Book)
Exploring Science Book 1
Exploring Science Book 2
Exploring Science Book 3
Exploring Science Book 4
Exploring Science Book 5
Exploring Science Book 6
Exploring Science Book 7
Exploring Science Book 8

**Focus On Series** unit study program — each title has a Student Textbook with accompanying Laboratory Notebook, Teacher's Manual, Lesson Plan, Study Notebook, Quizzes, and Graphics Package

Focus On Elementary Chemistry
Focus On Elementary Biology
Focus On Elementary Physics
Focus On Elementary Geology
Focus On Elementary Astronomy

Focus On Middle School Chemistry
Focus On Middle School Biology
Focus On Middle School Physics
Focus On Middle School Geology
Focus On Middle School Astronomy

Focus On High School Chemistry

**Super Simple Science Experiments**

21 Super Simple Chemistry Experiments
21 Super Simple Biology Experiments
21 Super Simple Physics Experiments
21 Super Simple Geology Experiments
21 Super Simple Astronomy Experiments
101 Super Simple Science Experiments

**Note:** A few titles may still be in production.

**Gravitas Publications Inc.**
www.gravitaspublications.com
www.realscience4kids.com

www.ingramcontent.com/pod-product-compliance
Lightning Source LLC
Chambersburg PA
CBHW051230200326
41519CB00025B/7317